D1487276

EXTREME CAREERS

OIL RIG WORKERS

Life Drilling for Oil

Katherine White

the rosen publishing group's
rosen central

Thank you Iain Bell and Mike Barnes for contributing time and expert information for this book

Published in 2003 by the Rosen Publishing Group, Inc.
29 East 21st Street, New York, NY 10010

Library of Congress Cataloging-in-Publication Data

White, Katherine.
Oil rig workers : life drilling for oil / Katherine White.— 1st ed.
    p. cm. — (Extreme careers)
Summary: Explores how to prepare for and get into the field of oil rig work as a leasehand, rigger, driller, or drilling manager, and looks at the daily life of one who chooses a career as an oil worker.
Includes bibliographical references and index.
ISBN 0-8239-3797-6 (lib. bdg.)
1. Oil well drilling—Vocational guidance—United States—Juvenile literature. [1. Oil well drilling—Vocational guidance. 2. Vocational guidance.] I. Title. II. Series.
TN871.2 .B55 2003
622'.3382'023—dc21

    2002008483

Manufactured in the United States of America

Contents

Introduction

Would you love to have a job where you travel all over the world? Does the idea of working only six months a year appeal to you? Do you love the outdoors and hope to find a career that allows you to work outside? These are some of the qualities of life as an oil rig worker!

Oil, also called petroleum, supplies 45 percent of the United States's energy needs. It is used in more than 6,000 different kinds of products. Highway pavement, car and airplane fuel, chewing gum, and cologne are all derived from oil. Oil is also used to heat most homes and businesses. Therefore, oil rig workers have very important jobs to do. They extract petroleum from deep crevices in the earth—and

Oil rig workers from Patterson Drilling Contractors put a drill pipe into the earth at a rig near Dew, Texas, on July 24, 2001.

petroleum is one of the most important fuels and ingredients used in the world today.

There are two basic types of oil drilling: onshore and offshore drilling. An oil rig is the large, complex machine designed to extract oil from the ground. When it is located in the water, the oil rig worker is doing offshore drilling. In contrast, onshore drilling is when the oil rig is built on land. Oil rig workers travel to the places where oil is found and live there while they work. If oil is found in the middle of the North Sea, the oil rig worker, or rigger, will go there to work. Most often, riggers only work for six months each year because while living on the oil rigs, they work many more hours per day than if they had office jobs. That's how huge an oil rig actually is—so big that hundreds of workers are able to live on it.

Along with the advantages of this exciting career, oil rig workers also face tough challenges every day. They work long hours on the rig and spend months away from their friends and families. Oil rig workers also face daily dangers, like blow outs. A blow out is an explosion that occurs when gas, oil, or salt water escapes in an uncontrolled manner from an oil well. So what makes oil rig workers love their jobs?

New Zealand native and twenty-year oil rig worker Jeremy Donaldson said it best in a recent article in the *Timaru Herald*: "I have friends all over the world and the chance to be part of the fascinating oil industry. It's a great life."

So how does traveling all over the world sound to you? Do you think you have what it takes to drill for oil? Launching a career as an oil worker is an exciting opportunity, one that brings many fascinating experiences. Read on to find out more about how to train to be an oil rig worker, and exactly what the life of an oil rig worker is like from early morning until night.

What Is an Oil Rig Worker?

An oil rig worker's job is to draw petroleum from the large oil reserves deep within the ground that formed millions of years ago. Riggers do not seek out the locations of such oil reserves themselves, as that is a whole other task in itself. Geologists and geo-physicists look for oil. Their jobs focus on identifying the deep crevices in the earth that hold petroleum. A rig worker concentrates on drilling the very deep hole that allows oil extraction, the process that actually draws the oil from the oil well. The huge drill located on the platform of the rig is an oil rig worker's most-used tool. In fact, without the drill, oil could not be pumped from the ground or from the bottom of the ocean.

Before rig workers can get started, scientists like geophysicist Stephen Lewis need to locate the oil. Here, he uses a 3-D virtual-reality computer model to explore reserves on the North Slope of Alaska.

Getting to Know the Drill

The derrick is one of the most essential parts of the rig. The derrick is a tall tower that raises and lowers all the drilling equipment, including the drill. The drill equipment makes the large hole in the ground or ocean floor. It looks a lot like a dentist's drill, except this drill is much bigger than the one that goes into your mouth. The drill has a sharp piece, called a rotary bit, or a bit, on one end that spins and grinds away the rock as the drill digs into the ground. The rotary bit has to cut away both soft and hard rock, so it is made of strong steel. The bit can dig up to 200 feet (61 meters) per hour when it's up against soft rock. Hard rock is another story: The bit might only cut 12 inches (30 centimeters) in an hour.

The rotary bit is attached to a section of the drill pipe, the tube that will carry the oil once it is found or as it is being pumped to the surface. The kelly is another section of the drill pipe that fits into a hole in the rotary table. The rotary table is turned by the rig's diesel engine as it works to rotate the drill pipe, kelly, and the rotary bit. Sound complicated? That's why the drilling rig manager must be an expert on everything that happens on the rig.

Drilling Manager: Expert on the Rig

An oil rig operates twenty-four hours a day, in eight-hour or longer shifts, with crews taking turns working

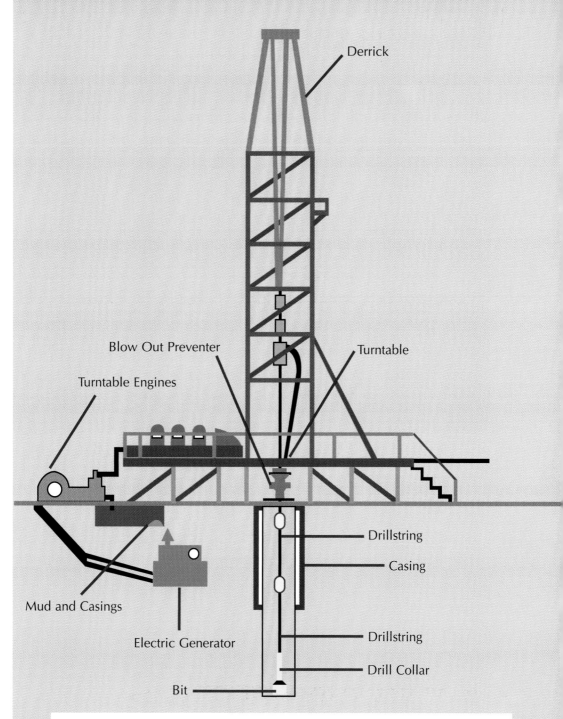

Derrick

Blow Out Preventer

Turntable

Turntable Engines

Drillstring

Casing

Mud and Casings

Drillstring

Electric Generator

Drill Collar

Bit

An oil rig is a powerful and complex piece of machinery that requires constant maintenance and skilled workers to operate it.

Where Do We Get Petroleum?

Petroleum was formed millions of years ago when most of the earth was covered by water. Over time, portions of the ocean dried up and the plants and animals that lived in these parts died. As they dried up, their remains slowly became fossils and were preserved in the earth. Sediments like clay, silt, and sand spread across the fossils. As the years passed, both the fossils and sediment became buried under many layers of rock. In fact, another name for petroleum is liquid rock. Deep in the earth, the layers did not receive oxygen, so they began to decompose together. As they decomposed, oil formed. The whole process took thousands of years.

the platform. The drilling manager, also called the toolpush, is in charge of his or her crew. This crew is made up of the driller, derrickhand, motorhand, and floorhand. Larger rigs or more complicated projects may have an additional floorhand, leasehand, or assistant driller.

The drilling manager is the person responsible for all drilling activity, meaning he or she supervises the team that actually does the drilling and works the derrick. However, the drilling manager also acts as the liaison, or the communicator, between the drilling company and the contractor. The following is a list of a drilling manager's official duties from the Alberta Learning Information Service in Canada (http://www.alis.gov.ab.ca):

- Coordinates the work for all crews
- Ensures that the entire drilling operation runs smoothly, safely, and efficiently
- Assures that operations comply with company policies and government regulations
- Makes arrangements to prepare areas for drilling
- Organizes the setting up and dismantling of rigs
- Ensures that emergency evacuation and medical procedures are posted and understood

That's a big job with a lot of responsibility, which is why the drilling manager is the highest rung on the ladder of rig workers. A drilling manager's most essential duty is to manage the crew. He or she oversees and interacts with the crew on a daily basis. This means that a drilling manager needs to have great communication skills. The drilling manager wants the platform to be a safe and satisfying place for the workers. This means he or she must also do a fair job when organizing duties. If you have great people skills, good coordination, and an ability to work safely and quickly, being a drilling manager might be a perfect position for you. And considering that drilling managers make between $75,000 and $100,000 per year, you'll be well paid for your efforts.

Becoming a Drilling Manager

A drilling manager must have at least two to three years of field experience—knowledge gained from direct work on rigs—on both onshore and offshore rigs. Working a number of years on a rig ensures a well-rounded background, a bonus both on and off the platform. A rigger needs more than just a positive attitude and field experience before becoming a toolpush, however. He or she must also attend classes to gain other necessary skills. The classes teach many safety and rig skills, educating the drilling manager on how to handle accidents and hazardous materials. Safety is the most important facet of working on a rig. The Alberta Learning

U.S. Navy lieutenant Doug Tunison *(left)* and Clarke Turner, director of the Naval Petroleum and Oil Shale Reserves in the Rocky Mountains, check an oil rig's progress.

Information Service recommends the following courses for those who want to become drilling managers.

Standard First-Aid Certificate

The drilling manager is trained to handle medical emergencies. Even during high-stress moments, the drilling manager must be able to handle the situation calmly and efficiently. All drilling managers must have a standard first aid certificate. In fact, everyone who works on a rig must have first-aid certification.

The certificate is earned when a person completes six and a half hours of first aid education. The course includes recognizing and caring for cardiac, or heart-related, emergencies in adults, identifying and caring for life-threatening bleeding, sudden illness, injuries, and heart disease prevention.

Transportation of Dangerous Goods (TDG) Training and Boiler Certificate

Safety regulations for oil rigs state: "No person shall handle, offer for transport, or transport dangerous goods, unless trained to do so." Therefore, oil rig workers must be taught how to handle dangerous materials, like gasoline, appropriately. Riggers learn that safety is the most important rule on a rig. Training for the boiler certificate teaches a rigger or drilling manager how to use and maintain a wide variety of boilers. The boiler is an important aspect of the rig and there are usually at least five on the platform. The boilers keep the oil at a safe temperature. By doing this, boilers prevent blow outs. The boiler certificate means the operator understands how the boiler works and how to manage it safely.

Workplace Hazardous Materials Information System (WHMIS) Training

Workplace Hazardous Materials Information System (WHMIS) training is designed to make sure the rig worker has the information he or she needs to work safely with hazardous materials. The class teaches safety concepts that will reduce illness and injury caused by hazardous materials in the workplace. The rigger also learns how to handle chemicals safely. Training stresses three key elements: Labels, Material Safety Data, and Sheets and Worker Education.

Second-Line Supervisors Well Control Certificate

The Petroleum Industry Training Service (PITS) offers a five-day course for the Second-Line Supervisors Well Control Certificate. The course deals with well control—checking oil temperature and machine levels—during open hole drilling operations. Participants are required to perform well control procedures and demonstrate the proper response to hole and equipment problems using test well equipment. The certificate earned is good for two years. After that time, the rig worker has to take the class again to get recertified.

Suitable Safety Training Related to Hydrogen Sulfide (H$_2$S)

This course covers the physical properties and health hazards of hydrogen sulfide, H$_2$S, a highly toxic chemical that is present in the crude oil and natural gas pumped through an oil well. Participants are required to operate self-contained breathing apparatus, use an H$_2$S detector device, and perform rescue breathing on a mannequin.

In high doses, inhalation of hydrogen sulfide can cause nausea and vomiting, coughing, confusion, and loss of consciousness. In extremely high doses, it can cause death.

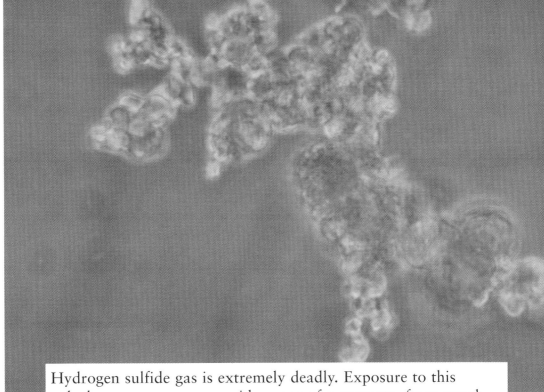

Hydrogen sulfide gas is extremely deadly. Exposure to this colorless gas can cause a wide range of symptoms, from coughing to loss of consciousness.

Awareness of Environmental Issues in the Drilling Industry

A drilling manager must be aware of all legislated rules and industry standards that apply to the work he or she is supervising. He or she must know the answers to any questions a crew may ask about safety codes, environmental impact, and work permits. This aspect of drilling manager training comes from being in touch and in tune with industry growth.

Growing into Rig Positions

As you may imagine, drilling managers have to make many tough decisions every day. So they must be experts—the best at everything that happens on a rig. Yet a drilling manager cannot be an expert without first going through a lot of learning experiences. That's why on a rig, riggers often work their way up a ladder of positions, depending on field experience and completed courses. The following chapter highlights the positions that a rigger usually works as his or her career grows on the rig.

Becoming a Rig Worker

Rig work is a ladder of opportunity. Each position on the rig requires different skill levels and backgrounds. They have different demands, tasks, and expectations. Each position also builds experience for a better position, the next rung of the rig work ladder. The complete ladder forms a rigging team. The drilling manager, as you learned in the first chapter, is at the top of that ladder, overseeing the entire crew. The next rung of the ladder is the driller, followed by the assistant driller, derrickhand, motorhand, floorhand, and finally the leasehand. All of the positions have certain perks that go along with the job.

Workers check on an oil rig in Texas in September 1998. Rig workers put in long hours for many consecutive days in a row.

Great Perks for Riggers

A rigger, no matter what his or her position on the rig, works seven twelve-hour shifts per week. These shifts include a paid meal and short breaks. On top of this, riggers work a fourteen/twenty-one day rotation. This means the rigger works for fourteen days straight and then has twenty-one days to rest and relax at home with his or her family. An employee's travel to and from home is often paid by the company. In addition, the company or contractor supplies its riggers with work clothes, housing, food, and laundry services when they are working on the rig.

Leasehand: An Entry-Level Rigger

The leasehand is the most basic position on the rig. This job is perfect for a person who does not have any experience on a rig but wants to get into rigging as a career. Rig work does not require a college degree, or even a high school diploma, but employers prefer that workers complete high school. Therefore,

World Oil Producers: A Breakdown

According to a 2001 survey by the Energy Information Administration, the following countries are the top oil producers:

Country	Production
Saudi Arabia	7.7 million barrels/day
Former Soviet Union	7.1 million barrels/day
United States	5.9 million barrels/day
Iran	3.6 million barrels/day
China	3.2 million barrels/day
Norway	3.0 million barrels/day
Mexico	3.0 million barrels/day
Venezuela	2.8 million barrels/day
United Kingdom	2.7 million barrels/day
Iraq	2.5 million barrels/day

rigging is a great career for individuals who choose not to attend college and want to start working right away. Finally, a person must be at least eighteen years old to work on a rig.

The leasehand position actually exists to help inexperienced individuals gain insight into and experience with all of the activities that go on around the rig. The leasehand performs general labor around the rig. This includes unloading trucks, digging ditches, building fences, and helping other crew members with maintenance.

The salary for a leasehand is between $750 to $1,100 per week, which equals around $110 to $150 per day. That's pretty good money for an entry-level position. As with all rig workers, the leasehand must have Workplace Hazardous Materials Information System (WHMIS) training and be certified in first aid.

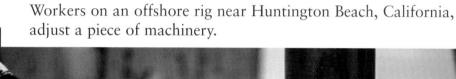

Workers on an offshore rig near Huntington Beach, California, adjust a piece of machinery.

Floorhand: Entry-Level Opportunity for Drilling

Since leasehands are on the bottom rung of the ladder, they often move up to the floorhand position rather quickly. Of course, the promotion has a lot to do with the leasehand's motivation and his or her ability to learn new tasks quickly. What are a floorhand's responsibilities?

The floorhand, also called the roughneck, spends most of the day working on the rig floor, scrubbing. The rig floor is the small work area where the rig crew conducts operations, usually adding or removing drill pipe to the drill. The floorhand, like the leasehand, helps other crew members in their duties. The floorhand also assists with maintenance and repair of rig equipment. Floorhands are a higher rung than the leasehand because they get more action and hands-on experience with the drill. The floorhand learns how to use wrench-like tongs, connect and disconnect pipe smoothly and efficiently, and assist in laying casing into a well after it has been drilled.

A floorhand usually earns around $310 to $345 per day or about $43,000 to $47,500 per year. A floorhand must also have first-aid certification, a WHMIS certificate, and an H_2S certificate.

Motorhand: Junior-Level Position

The motorhand, true to the position's name, operates and maintains the engines and all mechanical equipment on a rig. The motorhand plays a vital role because all the drills are usually run by three to four diesel engines and several electrical ones. Therefore, if the motorhand does not take good care of the engines, drilling productivity will decrease. Motorhands also have a managerial side to their position because they train leasehands and floorhands.

The salary is the same as a floorhand, around $43,000 to $47,500 per year, and the job requires the same training and certification.

Derrickhand: Living the High Life

The derrickhand is usually next in command to the driller in a basic rigging crew. Derrickhands spend about 20 percent of their time in the derrick, nine stories above the rig. The platform area from which the derrickhand works is called a monkeyboard. The rest of a derrickhand's time is spent maintaining drilling fluid. He or she controls the drilling fluid, or mud, and also works on fluid equipment.

The derrickhand position involves one very complicated task—using a harness and a platform climbing device. When the length of pipes on the drill, called the drillstring, is being raised or lowered, the derrickhand will be high in the air. Therefore, the derrickhand cannot be afraid of heights. Not everyone is comfortable being high above the ground. A person must work as a derrickhand, however, before he or she can be promoted to assistant driller or, even better, to the driller position.

The derrickhand earns between $40,000 to $50,000 per year, depending on experience and the company or contractor that hires him or her. Like the

Drill pipes on a derrick rise up from the earth in Weld County, Colorado. Derrickhands spend some of their time high on the derricks on a platform called the monkeyboard.

floorhand, the derrickhand must have standard first-aid certification, WHMIS training, and H_2S training.

Assistant Driller: Right-Hand Man or Woman

The assistant driller position, like the leasehand, exists as a training ground for the next rung of rig work. The job of the assistant driller, also called a roustabout, is to learn all of the duties of a driller.

An assistant driller works alongside the driller and assists the driller in his or her work. Assistant drillers are trained to operate the draw-works, rotary table, pumps, and drill pipe; keep a current record of the progress of the drilling operation; monitor the process of the drilling operation and determine needed changes; train crew members; and introduce new procedures. He or she also supervises the crew when the driller is too busy.

Assistant drillers require a great deal of training and certification, almost as much as a toolpush. They need training in standard first aid, H_2S, WHMIS, and

Workers at a derrick in Prudhoe Bay, Alaska, keep themselves warm against the cold weather, one of the perils of drilling in the far north.

First-Line Supervisors Blowout Prevention. Their average salary is around $310 to $345 dollars per day, which is around $42,000 to $46,000 per year.

Driller: Second in Charge

The driller position is one rung below the drilling manager or toolpush position. In fact, the driller reports directly to the drilling manager, providing

Drilling is hard, physical work that requires quick thinking and technical skills. Teamwork and a good chain of command are also valuable to the success of a rig.

daily reports and updates. The driller also has many other duties. In addition to being in charge of the operation of the rig and crew during their tour, drillers are in control of operating the drills and hoisting equipment and managing the rig floor. He or she is also in charge of the driller's console, which includes brakes, throttles, and clutches. The console is a machine that drillers use to make sure all temperature levels and drilling equipment are operating safely. The driller must be trained to read and operate the various gauges, checking for potential problems. Most important, the driller's console provides readouts, or information, about the drill's progress beneath the surface. By monitoring the console, the driller knows when the rig's gauges need to be changed and modified. An important part of the rig's operation, the driller needs at least two to three years of experience working on a rig. With two to five years of experience, a driller makes between $405 and $430 per day, which is about $54,000 to $58,000 per year. Just like all the other rig workers, the driller needs a standard first-aid certificate, First-Line Supervisors Blowout Prevention training, and an H_2S certificate.

Climbing to the Top

Now you know all the ins and outs of each worker's duties on the rig. Let's look at a day in the life of rig workers. What do riggers do with their free time? How often do they get to travel? What kind of food do rig workers eat and what kind of housing do they have? Riggers live on the rig, so it's really important to find out what their home away from home is really like.

A Day in the Life

Mike Barnes is the facilities supervisor on Kerr-McGee's Ninian Central Platform in the North Sea. The platform is halfway between the northern tip of the Shetland Islands and Norway. Mike's job is to oversee all of the non-drilling and oil production activities on the rig. He has a team of people who look after the cranes, lifecraft, accommodations, and supporting services for the rig workers—like catering, heating, fresh water, and fire pumps. Oil rig workers have jobs that are physically taxing. It's important to them that other aspects of life on the rig are enjoyable. Therefore, a big part of Mike's job is to make sure the rig is comfortable for the riggers.

Catering on a Rig?

Oil rigs have caterers who prepare all of the food for the riggers. The chefs are top-notch and work hard to make great dishes. They know how demanding rig work is and they want the riggers to start and finish their days with a good meal. Most of the time, the chefs prepare American meals with a lot of different meats, such as steak, roast beef, and chicken. These foods are full of protein, which is a great source of energy. Most dinners consist of meat, pasta, and salad.

Riggers also like to have a hearty breakfast because their work is so physically demanding. Fruit is a big hit in the morning, as are eggs and cereal. The catering chefs rotate in twelve-hour shifts just like the riggers, so there is always a way to get some food on a rig!

Mike's Day

As the facility supervisor, Mike wakes up at 4:00 AM. His shift officially runs from 7:00 AM to 7:00 PM. However, Mike chooses to be in the office by 5:00 AM each morning because it gives him a few hours to work before the platform comes alive at 7:00 AM. He explained in an interview: "This is my choice as I can get some of the time-consuming aspects of my paperwork done without the interruption of phone calls and visitors."

Mike has breakfast around 6:30 AM. He says the night-shift chef does not mind making him breakfast a little earlier than the rest of the crew. After breakfast, Mike heads to the control room area to sign permits. What sort of permits? As mentioned earlier, safety is incredibly important on a rig because the work is so dangerous. "Almost all activities on an oil platform have to be sanctioned by a permit to work. This is to ensure as much as possible that all safety aspects of the job have been considered."

At 7:00 AM Mike has his first meeting with the twenty people who make up the maintenance team; they discuss the day's activities. He then heads over to his 8:00 AM meeting. All of the major supervisors aboard the rig meet to discuss what their teams will be working on that day. This is to keep the lines of communications open and the platform safe. This is followed at 8:30 AM with a *conference call to the Kerr-McGee Aberdeen office to report any accidents or incidents, the major activities of the day, and achievements or failures of the previous day.*

Between the end of the conference call and lunch, Mike responds to e-mails, visits as many work sites as possible, and makes many phone calls to various engineers and people in the Aberdeen office. His afternoon is very similar as he continues his role of facilitating a wide variety of activities and work scopes as well as ensuring a safe and successful conclusion to each day. Though his evening meal is supposed to be at 7:00 PM, Mike usually has dinner at 8:00 PM because he likes to tidy his office for the start of another long, challenging day.

Safety on the Rig

Mike's job reflects how seriously safety is taken on a rig. Rig workers have a dangerous job because they constantly work with or alongside highly flammable fuels and very complex machinery. One flaw can

A group of workers at Murphy Oil lower a coworker in a gurney during a rescue and life-support skills competition at the Refinery Terminal Fire Company Academy in Corpus Christi, Texas.

cause a catastrophe. This is why safety permits and procedures are a focal point of the job. In fact, riggers try not to just think of offshore danger. Rather, they identify and consider the risks. Barnes says, "The most important tasks we carry out offshore are probably risk assessments on work activities, because if we incorrectly assess the risk factors, we could authorize an unsafe activity. Also the regular safety inspections and testing of safety devices and equipment are very important tasks."

All of their planning and decisions are based on the principle of ALARP—As Low As Reasonably Possible. "This means the riggers try to reduce all risks to ALARP by planning the methods we use to carry out the activities," explains Barnes.

Support Boats and Skilled Safety Workers

On top of risk assessment, all offshore rigs are circled by a support boat twenty-four hours a day. The boat regulates any dangerous activity on board, like fires. If a fire ignites, the support boat picks up the oil rig workers and carries them to safety.

Also, keep in mind that all rig workers are trained in first aid and a variety of other safety courses. This ensures that each worker knows how to deal with a potentially dangerous situation, as well as how to remain calm during an accident.

Safety is of the utmost importance on a rig and a lot of steps are taken to ensure a safe working environment. What happens when something goes wrong? What exactly are the dangers facing the riggers? The most catastrophic disaster is a blow out.

Accidents Happen: The World's Biggest Oil Rig Sinks

On March 15, 2001, disaster struck an oil rig off the coast of Brazil. Five days after explosions rocked the rig, called P-36, the forty-story structure sank into the ocean. Ten riggers were killed during the blast, which knocked over a support pillar. The lack of support caused part of the platform to lay in the ocean like a seesaw. For days, Petrobras, a major Brazilian oil company and owner of the rig, worked to salvage the damaged rig, but high winds and bad weather made it impossible.

Though rare, accidents do happen on oil rigs. Ten workers were killed by explosions on March 15, 2001, on P-36, the world's largest oil rig, off the shore of Macae, Brazil. Here, rescue teams attempt to salvage the rig, which began sinking after the explosions.

Blow Outs: Rare Dangers

A blow out occurs when an oil well becomes uncontrollable while drilling. At times, the spill can continue for hours, weeks, or even months. Blow outs cause explosions because the rig houses a number of flammable fuels and hot machinery. Fortunately, blow outs rarely occur. Approximately 1 percent of oil wells worldwide have had blows outs, which is an exceptionally low statistic. In fact, in more than 22,000 wells drilled in U.S. coastal waters from 1971 to 1993, only five blow outs have occurred.

Accommodations

Oil riggers are made to feel at home on the rig because they spend so much time away from their families. The cabins are small but designed to feel as comfortable as possible. The rooms usually have two bunks with a private shower and toilet facilities. Employers strive to give workers single occupancy, meaning that although rig workers have to share their quarters with someone else, they are bunked with someone who works a different shift. As a result, night shift workers often share with day shift workers.

Oil Rigger Hobbies

Oil rig workers spend a lot of time in the middle of the ocean or in fairly unpopulated areas of land, so they have to entertain themselves on the rig. What keeps them occupied? Most platforms have many different forms of entertainment for the riggers, such as a gym, a cinema, radios, saunas, a library, pool tables, a satellite TV, table tennis and other games, and a computer room. This keeps the workers entertained on a regular basis, giving them activities they can enjoy doing after a long shift on the platform.

Missing the Family

Mike Barnes loves his job—he's been doing it for almost twenty-three years—but he also acknowledges the drawbacks to being a rigger. "We spend more than 50 percent of our lives away from our families and loved ones. We are always offshore (according to our wives) when the car or washing machine breaks down. And, sometimes we get held on the platform beyond our trip due to bad weather. We often miss birthdays, anniversaries, school-prize days, and some of the significant milestones in our children's development."

Most rig workers agree that being away from their families is the hardest part about being a rigger. Nevertheless, they get used to it and so do their families. It's just one part of the job.

The Future of Rig Work

4

The leading producers of petroleum in the world are Russia, the United States, Saudi Arabia, Iran, Iraq, China, Mexico, the United Arab Emirates, Venezuela, Nigeria, Kuwait, and Norway. The largest petroleum reserves are in the Middle East. The location of the reserves explains why rig workers are scattered all over the world, in both the western and eastern hemispheres. Remember, rig workers go wherever oil is found.

Demand for Oil

The demand for oil is rising every year, in every country. The United States depends on oil for a substantial amount of its energy, around 45 percent. As of January 1993, the United States was using about 15.1 million barrels per day, which is around 634 million gallons. If lined up in one-gallon cans, they would encircle the earth's equator 4.5 times—about 117,000 miles of cans. That's a lot of oil used each and every day!

This is great for oil rig workers because their skills are in great demand. Countries and large oil companies, like ExxonMobil and ChevronTexaco, need riggers to drill for oil to meet rising oil demands. More oil

Rig worker Shane Battles readies a drilling pipe at a rig near Chickasha, Oklahoma, in December 2000. While the need for rig workers in certain regions may rise and fall, they will always be in demand as long as oil remains the lucrative market that it is.

wells will be found. More rigs will be built, needing more and more oil rig workers to drill for oil. This gives oil workers great job security. Companies need their unique skills.

Renewable Energy vs. Traditional Energies

Not everyone is pleased that the demand for oil is rising. Some scientists, for example, think the United States depends too much on oil. These scientists understand that oil is a fossil fuel and that it can be depleted. Oil, natural gas, and coal are all fossil fuels. Though fossil fuels form naturally in the earth, it takes thousands of years. Some scientists worry that a heavy dependence on oil will cause the reserves to dwindle. In other words, these fossil fuels will be used so quickly that new reserves will not form fast enough to replace them, and the world's oil supplies will be depleted.

This is where renewable energies come into the picture. Renewable energies also form naturally, but there is no way to deplete them. Examples of renewable energies are the Sun, wind, and water. You've most likely heard of solar energy and wind power—these are alternative technologies being used instead of fossil fuels.

The Environment

Fossil fuels such as oil also have a negative impact on the environment. When burned, these fuels release toxic gases, or emissions, into the air. These emissions harm the ozone layer and the environment, tampering with nature's balance and health. These issues are greatly debated because fossil fuels have been used for so long. Many people reject the idea that change is needed. Others work hard to show that change is necessary. Sometimes, riggers are caught in the middle.

On April 2, 2001, nine environmental activists took over a U.S. oil rig located in Inverness, Scotland. The activists were protesting global warming—a rise

George Bissell: Inventor of the First Oil Well

No one would have ever guessed that George Bissell would eventually become a mover and shaker in the Industrial Revolution. Independent at the age of twelve, he worked hard to put himself through Dartmouth College while teaching and writing articles. After graduation, he became a professor of Latin and Greek. He also worked as a journalist and even as a high school principal. After becoming ill in 1853, he went to Pennsylvania. There, he saw the inefficient way that oil was being gathered at the time. This memory stuck with him and when he returned to Dartmouth for a visit, he noticed a bottle of Pennsylvania rock oil used for healing. Knowing a thing or two about rocks and oil, he wondered if the substance could be used to light lanterns. He was right! But that was just the beginning. First he had to figure out a better way to extract oil from the ground. Taking information from the salt drilling industry and teaming up with Colonel Edwin L. Drake, Bissell eventually found success after a long and weary time. The two struck oil on August 28, 1859, becoming the legends they are today.

Environmental activists on board this Greenpeace ship, the MV *Arctic Sunrise*, blocked Atlantic Richfield's drilling platform in the Beaufort Sea on August 13, 1997, to protest the continuing reliance of the West on fossil fuels, which they see as dangerous.

in the earth's temperature that harms the environment. The Greenpeace environmentalists objected to U.S. president George W. Bush's decision to drop a 1997 treaty aimed at curbing greenhouse gases believed by many scientists to create global warming. Others are concerned that drilling for oil on land will disturb wildlife and pollute soil and water. Certain areas, such as the Arctic, are home to many endangered animals. Conservationists are dedicated to preserving such areas.

The debate continues to be a heated one, with many diverse opinions and a lot of money fueling the debate. Riggers try not to get involved in this arena, because they are just doing their jobs. They work hard to carry out each task safely and to protect the environment. Although drilling for oil is controversial, as long as people are dependent on it for fuel and other things, the quest for oil will continue.

Expectations and Outlook

In the meantime, riggers continue to do what they know best: drill and pump oil. This is their job, and

Oil Rig Workers: Life Drilling for Oil

for some oil rig workers, this is their passion. A career in rig work is filled with adventure and action. A day's work is long and full of important tasks and decisions. Most rig workers love what they do. They enjoy traveling all over the world and meeting new people. They love drilling for oil and feeling the excitement that surrounds searching for such an important fuel.

Glossary

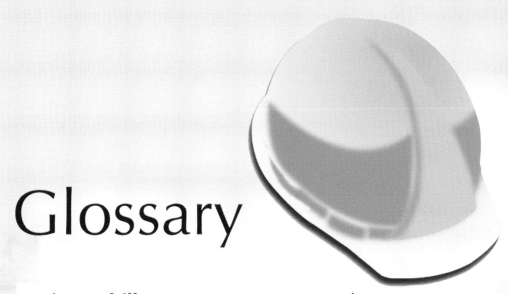

assistant driller A training position to become a driller.

controversial An issue that is highly debated; having opposing viewpoints.

crevice A small crack or narrow opening.

crude A substance in its natural state; an unprocessed material.

decompose A process in which something decays or breaks down into simpler compounds.

deplete To empty; a reduction in quantity.

derrick A tall tower over an oil well that raises and lowers all the drilling equipment and the drill.

derrickhand A person who spends 20 percent of the time nine stories above an oil well platform on the derrick and the rest of the time maintaining drill fluid.

driller The second person in charge of a rig below the toolpush; the driller oversees the crew and handles most drilling activity.

drilling manager The person in charge of an entire rig crew; also called a toolpush.

emission A substance released into the air.

floorhand An entry-level position on an oil rig; floorhands learn how to place drill pipe on the drillstring and assist most of the crew.

fossil fuels Naturally occurring fuels that are formed within the earth from decomposing plant and animal remains; includes coal, oil, and natural gas.

global warming A rise in the earth's temperature that harms the environment.

inhalation The act of breathing in.

leasehand The starting entry-level position on an oil rig; leasehands help unload trucks, dig ditches, and perform any basic task a rigger commands.

motorhand A person who operates and maintains the engines and all mechanical equipment on a rig; also trains the leasehands and floorhands.

petroleum Fossil fuel that is formed in the ground over thousands of years; the fuel for which riggers drill.

platform area An area on an oil rig above ground where the drill and derrick are located, and on which rig workers often do their jobs.

rotary bit The piece of machinery that spins and grinds away rock as an oil rig drill digs into the ground; also called the bit.

For More Information

American Association of Drilling Engineers
2301 Belmont Place
Metairie, LA 70001
(707) 504-4177
e-mail: info@aade.org
Web site: http://www.aade.org

American Petroleum Institute
1220 L Street NW
Washington, DC 20005-4070
(202) 682-8000
Web site: http://www.api.org

CAZA Safety Plus and RIG PASS Program
CAZA Drilling, Inc.
1801 Broadway, Suite 360

Denver, CO 80202
(303) 292-1206
e-mail: dan@cazadrilling.com
Web site: http://www.cazadrilling.com

In Canada

Canadian Association of Oilwell Drilling Contractors
#800, 540-5 Avenue SW
Calgary, AB T2P 0M2
(403) 264-4311
e-mail: info@caodc.ca
Web site: http://www.caodc.ca/genpub.htm

Canadian Association of Petroleum Producers
350 7th Avenue SW, Suite 2100
Calgary, AB T2P 3N9
e-mail: communication@capp.ca
Web site: http://www.capp.ca

Maritime Drilling Schools Ltd.
105 Regent Street
P.O. Box 1916
North Sydney, NS B2A 3S9

(866) 807-3960
e-mail: mds@ns.sympatico.ca
Web site: http://www.mdslimited.ca

Petroleum Industry Training Service (PITS)
Nisku Training Centre
1020-20th Avenue
Nisku, AB T9E 7Z6
(800) 387-4976
Web site: http://www.pits.ca

In the United Kingdom

Offshore Contractors Association
58 Queens Road
Aberdeen, Scotland AB15 4YE
United Kingdom
+44 1224 326070
e-mail: info@ukooa.co.uk
Web site: http://www.oca-online.co.uk

Web Sites

Due to the changing nature of Internet links, the Rosen Publishing Group, Inc., has developed an online list of Web sites related to the subject of this book. This site is updated regularly. Please use this link to access the list:

http://www.rosenlinks.com/ec/oirw/

For Further Reading

Aaseng, Nathan. *Business Builders in Oil.* Minneapolis, MN: Oliver Press, 2000.

Alcraft, Rob. *Oil Disasters.* Des Plaines, IL: Heinemann Library, 2000.

Conaway, Charles F. *The Petroleum Industry: A Nontechnical Guide.* Tulsa, OK: PennWell Books, 1999.

Dineen, Jacqueline. *Oil, Gas, and Coal.* New York: Raintree Steck-Vaughn Publishers, 1995.

Bibliography

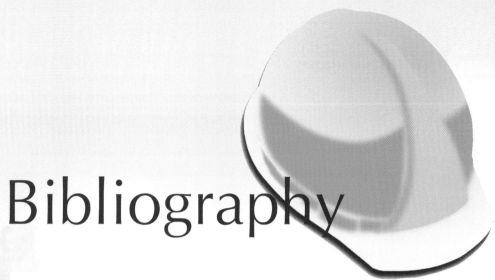

Ardley, Neil. *How We Build: Oil Rigs.* Ada, OK: Garrett Educational Corporation, 1990.

CNN. "Oil Rig Occupation Continues." April 2, 2001. Retrieved January 5, 2002 (http://www.cnn.com/2001/WORLD/europe/04/02/scotland.protest02).

CNN. "Worlds Biggest Oil Rig Sinks." March 20, 2001. Retrieved January 5, 2002 (http://www.cnn.com/2001/WORLD/americas/03/20/brazil.rig.02).

Lynch, Michael. *How Oil Rigs Are Made.* New York: Facts on File, 1985.

Petroleum Industry Training Service. "Training Main Menu." Retrieved January 5, 2002 (http://www.pits.ca/crsmenu.html).

Index

About the Author

Katherine White is a freelance writer and editor. She lives in Jersey City, New Jersey.

Photo Credits

Cover © Joe Baraban/Corbis; pp. 5, 9, 15, 38, 41, 46, 50 © AP/Wide World Photos; p. 19 © Lester V. Bergman/Corbis; p. 22 © John B. Boykin/Corbis; p. 25 © Vince Streano/Corbis; p. 29 © Lowell Georgia/Corbis; p. 31 © James A. Sugar/Corbis; p. 32 © Lester Lefkowitz/Corbis.

Editor

Eliza Berkowitz

Series Design

Les Kanturek

Layout

Tahara Hasan